The Adductor Muscles of the Jaw In Some Primitive Reptiles

Richard C. Fox

Classical Prints

This Edition Published in 2022

ISBN: 9789356536968

TABLE OF CONTENTS

Information about osteological changes in the groups of reptiles that gave rise to mammals is preserved in the fossil record, but the musculature of these reptiles has been lost forever. Nevertheless, a reasonably accurate picture of the morphology and the spatial relationships of the muscles of many of these extinct vertebrates can be inferred by studying the scars or other marks delimiting the origins and insertions of muscles on the skeletons of the fossils and by studying the anatomy of Recent genera. A reconstruction built by these methods is largely speculative, especially when the fossil groups are far removed in time, kinship and morphology from Recent kinds, and when distortion, crushing, fragmentation and overzealous preparation have damaged the surfaces associated with the attachment of muscles. The frequent inadequacy of such direct evidence can be partially offset by considering the mechanical demands that groups of muscles must meet to perform a particular movement of a skeletal member.

Both direct anatomical evidence and inferred functional relations were used to satisfy the purposes of the study here reported on. The following account reports the results of my efforts to: 1, reconstruct the adductor muscles of the mandible in *Captorhinus* and *Dimetrodon*; 2, reconstruct the external adductors of the mandible in the cynodont *Thrinaxodon*; and 3, learn the causes of the appearance and continued expansion of the temporal fenestrae among the reptilian ancestors of mammals.

The osteology of these three genera is comparatively well-known. Although each of the genera is somewhat specialized, none seems to have departed radically from its relatives that comprised the line leading to mammals.

I thank Prof. Theodore H. Eaton, Jr., for suggesting the study here reported on, for his perceptive criticisms regarding it, and for his continued patience throughout my investigation. Financial assistance was furnished by his National Science Foundation Grant (NSF-G8624) for which I am also appreciative. I thank Dr. Rainer Zangerl, Chief Curator of Geology, Chicago Museum of Natural History, for permission to examine the specimens of *Captorhinus* and *Dimetrodon* in that institution. I am grateful to Mr. Robert F. Clarke, Assistant Professor of Biology, The Kansas State Teachers College, Emporia, Kansas, for the opportunity to study his specimens of *Captorhinus* from Richard's Spur, Oklahoma. Special acknowledgment is due Mr. Merton C. Bowman for his able preparation of the illustrations.

Captorhinus

The outlines of the skulls of *Captorhinus* differ considerably from those of the skulls of the primitive captorhinomorph *Protorothyris*. Watson (1954:335, Fig. 9) has shown that in the morphological sequence, *Protorothyris—Romeria—Captorhinus*, there has been flattening and rounding of the skull-roof and loss of the primitive "square-cut" appearance in transverse section. The quadrates in *Captorhinus* are farther from the midline than in *Protorothyris*, and the adductor chambers in *Captorhinus* are considerably wider than they were primitively. Additionally, the postorbital region of *Captorhinus* is relatively longer than that of *Protorothyris*, a specialization that has increased the length of the chambers within.

In contrast with these dimensional changes there has been little shift in the pattern of the dermal bones that roof the adductor chambers. The most conspicuous modification in *Captorhinus* is the absence of the tabular. This element in *Protorothyris* was limited to the occiput and rested without sutural attachment upon the squamosal (Watson, 1954:338); later loss of the tabular could have had no effect upon the origins of muscles from inside the skull roof. Changes in pattern that may have modified the origin of the adductors in *Captorhinus* were correlated with the increase in length of the parietals and the reduction of the supratemporals. Other changes that were related to the departure from the primitive romeriid condition of the adductors included the development of a coronoid process, the flattening of the quadrate-articular joint, and the development of the peculiar dentition of *Captorhinus*.

The adductor chambers of *Captorhinus* are large. They are covered dorsally and laterally by the parietal, squamosal, postfrontal, postorbital, quadratojugal and jugal bones. The chamber extends medially to the braincase, but is not limited anteriorly by a bony wall. The occiput provides the posterior limit. The greater part of the adductor chambers lies mediad of the mandibles and thus of the Meckelian fossae; consequently the muscles that arise from the dermal roof pass downward and outward to their insertion on the mandibular rami.

Mandible

The mandibular rami of *Captorhinus* are strongly constructed. Each ramus is slightly convex in lateral outline. Approximately the anterior half of each ramus lies beneath the tooth-row. This half is roughly wedge-shaped in its lateral aspect, reaching its greatest height beneath the short posterior teeth.

The posterior half of each ramus is not directly involved in supporting the teeth, but is associated with the adductor musculature and the articulation of the ramus with the quadrate. The ventral margin of this part of the ramus curves dorsally in a gentle arc that terminates posteriorly at the base of the retroarticular process. The dorsal margin in contrast sweeps sharply upward behind the teeth and continues posteriorly in a long, low, truncated coronoid process.

A prominent coronoid process is not found among the more primitive members of the suborder, such as *Limnoscelis*, although the mandible commonly curves upward behind the tooth-row in that genus. This area in *Limnoscelis* is overlapped by the cheek when the jaw is fully adducted (Romer, 1956:494, Fig. 213), thereby foreshadowing the more extreme condition in *Captorhinus*.

The coronoid process in *Captorhinus* is not oriented vertically, but slopes inward toward the midline at approximately 45 degrees, effectively roofing the Meckelian fossa and limiting its opening to the median surface of each ramus. When the jaw was adducted, the coronoid process moved upward and inside the cheek. A space persisted between the process and the cheek because the process sloped obliquely away from the cheek and toward the midline of the skull. The external surface of the process presented an area of attachment for muscles arising from the apposing internal surface of the cheek.

Palate

The palate of *Captorhinus* is of the generalized rhynchocephalian type (Romer, 1956:71). In *Captorhinus* the pterygoids and palatines are markedly arched and the relatively large pterygoid flange lies almost entirely below the lower border of the cheek. The lateral edge of the flange passes obliquely across the anterior lip of the Meckelian fossa and abuts against the bottom lip of the fossa when the jaw is closed.

The palatines articulate laterally with the maxillary bones by means of a groove that fits over a maxillary ridge. This presumably allowed the halves of the palate to move up and down rather freely. The greatest amplitude of movement was at the midline. Anteroposterior sliding of the palate seems impossible in view of the firm palatoquadrate and quadrate-quadratojugal articulations.

The subtemporal fossa is essentially triangular, and its broad end is bounded anteriorly by the pterygoid flange. The fossa is lateral to much of the adductor chamber; consequently muscles arising from the parietals passed ventrolaterally, parallel to the oblique quadrate ramus of the pterygoid, to their attachment on the mandible.

Musculature

These osteological features indicate that the adductor muscles of the jaw in *Captorhinus* consisted of two primary masses (Figs. 1, 2, 3). The first of these, the capitimandibularis, arose from the internal surface of the cheek and roof of the skull and inserted on the bones of the lower jaw that form the Meckelian canal and the coronoid process.

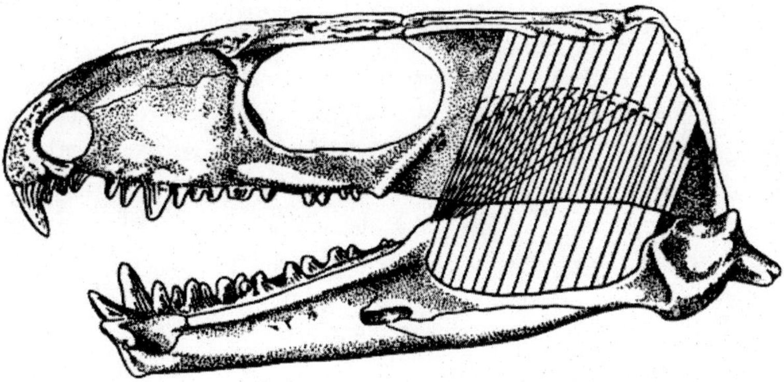

Fig. 1. Captorhinus. Internal aspect of skull, showing masseter, medial adductor, and temporal muscles. Unnumbered specimen, coll. of Robert F. Clarke. Richard's Spur, Oklahoma. × 2.

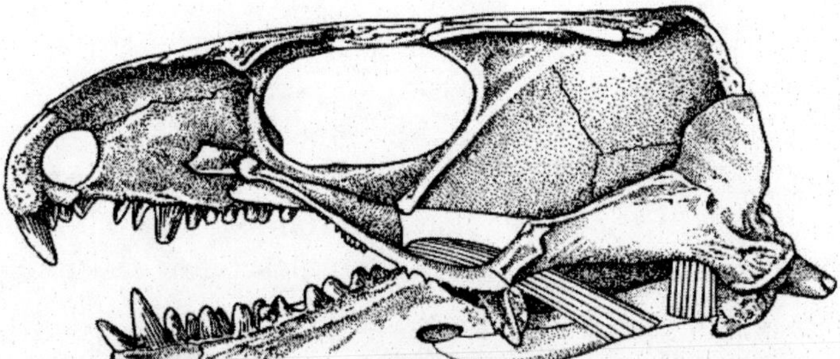

Fig. 2. Captorhinus. Internal aspect of skull, showing anterior and posterior pterygoid muscles. Same specimen shown in Fig. 1. × 2.

The muscle was probably divided into a major medial mass, the temporal, and a lesser, sheetlike lateral mass, the masseter. The temporal was the largest of the adductors and arose from the lateral parts of the parietal, the dorsal parts of the postorbital, the most posterior extent of the postfrontal, and the upper parts of the squamosal. The muscle may have been further subdivided, but evidence for subordinate slips is lacking. The fibers of this mass were nearly vertically oriented in lateral aspect since the parts of the ramus that are available for their insertion lie within the anteroposterior extent of the adductor chamber. In anterior aspect the fibers were obliquely oriented, since the jaw and subtemporal fossa are lateral to much of the skull-roof from which the fibers arose.

The masseter probably arose from the quadratojugal, the jugal, and ventral parts of the squamosal, although scars on the quadratojugal and jugal are lacking. The squamosal bears an indistinct, gently curved ridge, passing upward and forward from the posteroventral corner of the bone and paralleling the articulation of the squamosal with the parietal. This ridge presumably marks the upper limits of the origin of the masseter from the squamosal.

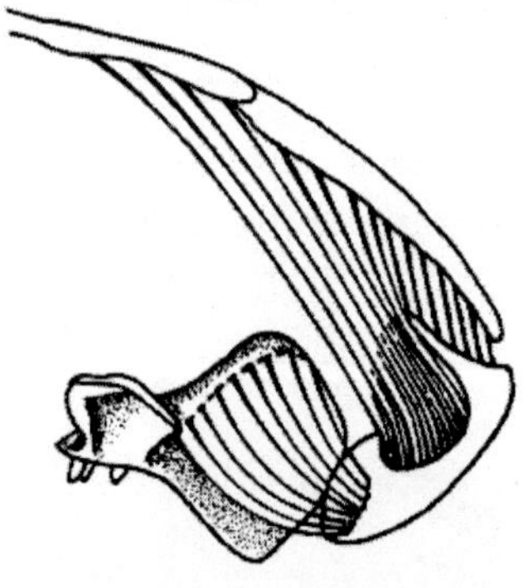

Fig. 3. Captorhinus. Cross-section of right half of skull immediately behind the pterygoid flange, showing masseter, temporal, and anterior pterygoid muscles. Same specimen shown in Fig. 1. × 2.

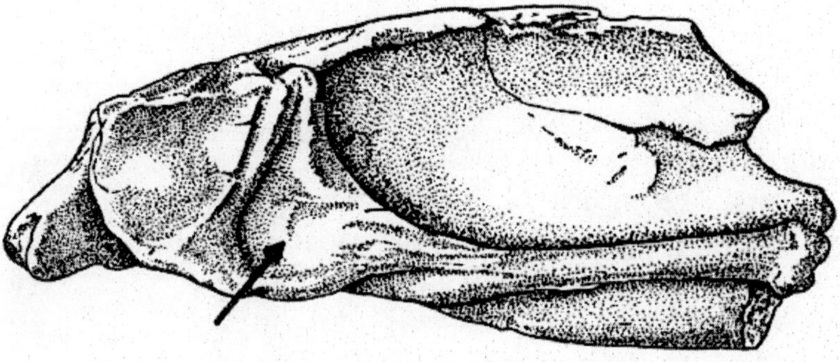

Fig. 4. Captorhinus. Internal aspect of left mandibular fragment, showing insertion of posterior pterygoid muscle. KU 8963, Richard's Spur, Oklahoma. × 2.8.

The masseter inserted on the external surface of the coronoid process, within two shallow concavities separated by an oblique ridge. The concavities and ridge may indicate that the muscle was divided into two sheets. If so, the anterior component was wedge-shaped in cross-section, and its thin posterior edge overlapped the larger mass that inserted on the posterior half of the coronoid process.

From a functional standpoint it is doubtful that a major component of the adductors arose from the quadrate wing of the pterygoid, for when the jaw is closed the Meckelian fossa is directly lateral to that bone. If the jaw were at almost any angle but maximum depression, the greatest component of force would be mediad, pulling the rami together and not upward. The mediad component would increase as the jaw approached full adduction. Neither is there anatomical evidence for an adductor arising from the quadrate wing of the pterygoid. The bone is smooth, hard, and without any marks that might be interpreted as muscle scars.

The internal adductor or pterygoid musculature in *Captorhinus* consisted of anterior and posterior components. The anterior pterygoid arose from the lateral edge and the dorsal surface of the pterygoid flange. The burred dorsal recurvature of the edge resembles that of the flange of crocodiles, which serves as part of the origin of the anterior pterygoid in those animals. In *Captorhinus* the attachment of the anterior pterygoid to the edge of the flange was probably tendinous, judging from the extent of the development of the edge of the flange. From the edge the origin extended medially across the dorsal surface of the flange; the ridging of this surface is indistinct, leading to the supposition that here the origin was more likely to have been fleshy than tendinous.

The anterior pterygoid extended obliquely backward and downward from its origin, passed medial to the temporal muscle and inserted on the ventral and medial surfaces of the splenial and angular bones beneath the Meckelian fossa. The spatial relationship between the palate and quadrate-articular joint indicate that the muscle was probably a minor adductor in *Captorhinus*.

When the jaw was adducted, the insertion of the anterior pterygoid was in a plane nearly level with the origin. Contraction of the anterior pterygoid when the jaw was in this position pulled the mandible forward and did not adduct it. Maximum depression of the mandible produced maximum disparity vertically between the levels of the origin and insertion. The force exerted by the anterior pterygoid upon the mandible when fully lowered most nearly approached the perpendicular to the long axes of the mandibular rami, and the resultant force acting on the mandible was adductive.

The adductive component of force therefore decreased as the jaw swung upward, with the result that the anterior pterygoid could only have been active in initiating adduction and not in sustaining it.

The evidence regarding the position and extent of the posterior pterygoid is more veiled. On the medial surface of the mandible, the prearticular and articular bones meet in a ridge that ventrally rims the glenoid cavity (Fig. 4). The ridge extends anteriorly and curves slightly in a dorsal direction and meets the Meckelian fossa. The curved part of the ridge is made of the prearticular bone alone. A small hollow above the ridge, anterior to the glenoid cavity, faces the medial plane of the skull and is bordered by the articular bone behind and above, and by the Meckelian fossa in front.

The surfaces of the hollow and the prearticular-articular ridge bear tiny grooves and ridges that seem to be muscle scars. The entire area of the hollow and its bordering features was probably the area of insertion of the posterior pterygoid.

However, the area of insertion lies mostly ventral to the articulating surface of the articular bone and extends but slightly in front of it. Seemingly little lever effect could be exercised by an adductor attaching in this position, namely, at the level of the fulcrum of the mandibular ramus.

The posterior pterygoid muscle probably arose from the anterior portion of the pterygoid wing of the quadrate, from a ridge on the ventromedial surface. From the relationship of the muscle to the articulation of the jaw with the skull, it may be deduced that the muscle was limited in function to the stabilization of the quadrate-articular joint by keeping the articular surfaces in close contact with each other and by preventing lateral slipping.

Finally there is evidence for an adductor between the temporal and masseter masses. The anterior dorsal lip of the Meckelian fossa supports a small knob to which this muscle attached, much as in *Sphenodon* (Romer, 1956:18, Fig. 12). Presumably the muscle was sheetlike and attached to the skull roof, medial to the attachment of the masseter.

A pseudotemporal may have been present, but evidence to indicate its extent and position is lacking. The muscle usually arises from the epipterygoid and nearby areas of the braincase and skull roof and inserts in the anterior parts of the fossa of the jaw. In *Captorhinus* the lateral wing of the pterygoid cuts across the fossa, effectively blocking it from the upper and medial parts of the skull, the areas of origin for the pseudotemporal.

Dimetrodon

The morphology of the skull of *Dimetrodon* closely resembles that of the primitive *Haptodus* (Haptodontinae, Sphenacodontidae), and "hence may be rather confidently described as that of the family as a whole" (Romer and Price, 1940:285). The major differences between the two genera are in the increased specialization of the dentition, the shortening of the lacrimal, and the development of long vertebral spines in *Dimetrodon*. The absence of gross differences in the areas of the skull associated with the groups of muscles with which this study is concerned, implies a similarity in the patterns of musculature between the two groups. Romer and Price suggest that *Haptodus*, although too late in time to be an actual ancestor, shows "all the common features of the *Dimetrodon* group on the one hand and the therapsids on the other." The adductors of the jaw of *Dimetrodon* were probably little changed from those of the Haptodontinae and represent a primitive condition within the suborder.

Dimetrodon and *Captorhinus* differ in the bones associated with the adductor mechanism; the area behind the orbit in *Dimetrodon* is relatively shorter, reducing the comparative longitudinal extent of the adductor chamber. Furthermore, the dermal roof above the adductor chamber slopes gently downward from behind the orbit to its contact with the occipital plate in *Dimetrodon*. Temporal fenestrae are, of course, present in *Dimetrodon*.

Musculature

The adductor musculature of the lower jaw in *Dimetrodon* was divided into lateral and medial groups (Figs. 5, 6). The lateral division consisted of temporal and masseter masses. The temporal arose from the upper rim of the temporal opening, from the lateral wall of the skull behind the postorbital strut, and from the dorsal roof of the skull. The bones of origin included jugal, postorbital, postfrontal, parietal and squamosal. This division may also have arisen from the fascia covering the temporal opening (Romer and Price, 1940:53). The muscle passed into the Meckelian fossa of the mandible and inserted on the angular, surangular, prearticular, coronoid and dentary bones. Insertion on the lips of the fossa also probably occurred.

The lateral division arose from the lower rim of the temporal opening and from the bones beneath. Insertion was in the Meckelian fossa and on the dorsal surface of the adjoining coronoid process.

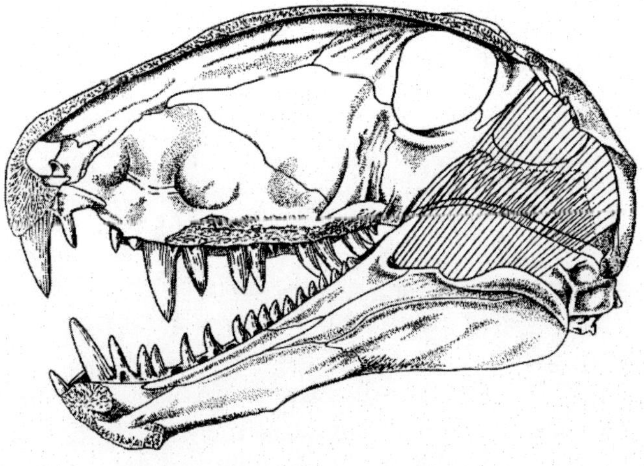

Fig. 5. Dimetrodon. Internal aspect of skull, showing masseter and temporal muscles. Skull modified from Romer and Price (1940). Approx. × 1/4.

The reconstruction of the progressively widening masseter as it traveled to the mandible follows from the progressively widening depression on the internal wall of the cheek against which the muscle must have been appressed. The depressed surface included the posterior wing of the jugal, the whole of the squamosal, and probably the anteriormost parts of the quadratojugal. Expansion of the muscle rostrally was prevented by the postorbital strut that protected the orbit (Romer and Price, 1940:53).

The sphenacodonts possess the primitive rhynchocephalian kind of palate. In *Sphenodon* the anterior pterygoid muscle arises from the dorsal surface of the pterygoid bone and from the adjacent bones. A similar origin suggests itself for the corresponding muscle, the second major adductor mass, in *Dimetrodon*.

From the origin the muscle passed posterodorsad and laterad of the pterygoid flange. Insertion was in the notch formed by the reflected lamina of the angular, as suggested by Watson (1948).

In *Dimetrodon* the relationship of the dorsal surface of the palate and the ventromedial surface of the mandible in front of the articulation with the quadrate is unlike that in *Captorhinus*. When the mandible of *Dimetrodon* is at rest (adducted), a line drawn between these two areas is oblique, between 30 and 40 degrees from the horizontal. Depression of the mandible increases this angle. The insertion of the anterior pterygoid is thus always considerably below the origin, permitting the muscle to be active throughout the movement of the mandible, from maximum depression to complete adduction. This was a major factor in adding substantially to the speed and power of the bite.

The presence and extent of a posterior pterygoid is more difficult to assess, because of the closeness of the glenoid cavity and the raised ridge of the prearticular, and the occupancy of at least part of this region by the anterior pterygoid. In some specimens of *Dimetrodon* the internal process of the articular is double (see Romer and Price, 1940:87, Fig. 16) indicating that there was a double insertion here. Whether the double insertion implies the insertion of two separate muscles is, of course, the problem. Division of the pterygoid into anterior and posterior portions is the reptilian pattern (Adams, 1919), and such is adhered to here, with the posterior pterygoid arising as a thin sheet from the quadrate wing of the pterygoid and the quadrate, and inserting by means of a tendon on the internal process of the articular, next to the insertion of the anterior pterygoid.

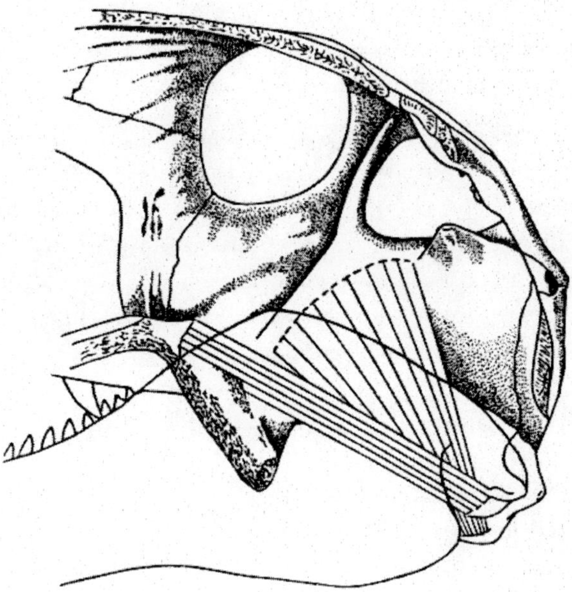

Fig. 6. Dimetrodon. Internal aspect of right cheek, showing anterior and posterior pterygoid muscles. Skull modified from Romer and Price (1940). Approx. × 1/4.

Watson (1948) has reconstructed the musculature of the jaw in *Dimetrodon* with results that are at variance with those of the present study. Watson recognized two divisions, an inner temporal and an outer masseteric, of the capitimandibularis, but has pictured them (830: Fig. 4; 831: Fig. 5C) as both arising from the inner surface of the skull roof above the temporal opening. But in *Captorhinus* the masseter arose from the lower part of the cheek close to the outer surface of the coronoid process. Watson has shown (1948:860, Fig. 17B) the same relationship of muscle to zygoma in *Kannemeyeria sp.* It is this arrangement that is also characteristic of mammals and presumably of *Thrinaxodon*. In view of the consistency of this pattern, I have reconstructed the masseter as arising from the lower wall of the cheek beneath the temporal opening.

Watson's reconstruction shows both the temporal and masseter muscles as being limited anteroposteriorly to an extent only slightly greater than the anteroposterior diameter of the temporal opening. The whole of the posterior half of the adductor chamber is unoccupied. More probably this area was filled by muscles. The impress on the inner surface of the cheek is evident, and the extent of both the coronoid process and Meckelian opening beneath the rear part of the chamber indicate that muscles passed through this area.

Watson remarked (1948:829-830) that the Meckelian opening in *Dimetrodon* "is very narrow and the jaw cavity is very small. None the less, it may have been occupied by the muscle or a ligament connected to it. Such an insertion leaves unexplained the great dorsal production of the dentary, surangular and coronoid. This may merely be a device to provide great dorsal-ventral stiffness to the long jaw, but it is possible and probable that some part of the temporal muscle was inserted on the inner surface of the coronoid. Indeed a very well-preserved jaw of *D. limbatus?* (R. 105: Pl. I, Fig. 2) bears a special depressed area on the outer surface of the extreme hinder end of the dentary which differs in surface modelling from the rest of the surface of the jaw, has a definite limit anteriorly, and may represent a muscle insertion. The nature of these insertions suggests that the muscle was already divided into two parts, an outer masseter and an inner temporalis." But, unaccountably, Watson's illustration (1948:830, Fig. 4) of his reconstruction limits the insertion of the temporal to the anterior limit of the Meckelian opening and a part of the coronoid process above it. No muscle is shown entering the Meckelian canal. It seems more likely that the temporal entered and inserted in the canal and on its dorsal lips. The masseter inserted lateral to it, over the peak of the coronoid process, and overlapping onto the dorsalmost portions of its external face, as Watson has illustrated (Plate I, middle fig.).

I am in agreement with Watson's reconstruction of the origins for both the anterior and posterior pterygoid muscles. On a functional basis, however, I would modify slightly Watson's placement of the insertions of these muscles. Watson believed that the jaw of *Dimetrodon* was capable of anteroposterior sliding. The articular surfaces of the jaws of *Dimetrodon* that I have examined indicate that this capability, if present at all, was surely of a very limited degree, and in no way comparable to that of *Captorhinus*. The dentition of *Dimetrodon* further substantiates the movement of the jaw in a simple up and down direction. The teeth of *Dimetrodon* are clearly stabbing devices; they are not modified at all for grinding and the correlative freedom of movement of the jaw that that function requires in an animal such as *Edaphosaurus*. Nor are they modified to parallel the teeth of *Captorhinus*. The latter's diet is less certain, but presumably it was insectivorous (Romer, 1928). With the requisite difference in levels of origin and insertion of the anterior pterygoid in *Dimetrodon* insuring the application of force throughout the adduction of the jaws, it would seem that the whole of the insertion should be shifted downward and outward in the notch. If this change were made in the reconstruction, the anterior pterygoid would have to be thought of as having arisen by a tendon from the ridge that Watson has pictured (1948:828, Fig. 3) as separating his origins for anterior and posterior pterygoids. The posterior pterygoid, in turn, arose by tendons from the adjoining lateral ridge and from the pterygoid process of Romer and Price. Tendinous origins are indicated by the limitations of space in this area, by the strength of the ridges pictured and reported by Watson, and by the massiveness of the pterygoid process of Romer and Price.

Discussion

A comparison of the general pattern of the adductor musculature of *Captorhinus* and *Dimetrodon* reveals an expected similarity. The evidence indicates that the lateral and medial temporal masses were present in both genera. The anterior pterygoid aided in initiating adduction in *Captorhinus*, whereas in *Dimetrodon* this muscle was adductive throughout the swing of the jaw. Evidence for the presence and extent of a pseudotemporal muscle in both *Captorhinus* and *Dimetrodon* is lacking. The posterior division of the pterygoid is small in *Captorhinus*. In *Dimetrodon* this muscle has been reconstructed by Watson as a major adductor, an arrangement that is adhered to here with but slight modification.

The dentition of *Captorhinus* suggests that the jaw movement in feeding was more complex than the simple depression and adduction that was probably characteristic of *Dimetrodon* and supports the osteological evidence for a relatively complex adductor mechanism.

In *Captorhinus* the presence of an overlapping premaxillary beak bearing teeth that are slanted posteriorly requires that the mandible be drawn back in order to be depressed. Conversely, during closure, the jaw must be pulled forward to complete full adduction. The quadrate-articular joint is flat enough to permit such anteroposterior sliding movements. The relationship of the origin and insertion of the anterior pterygoid indicates that this muscle, ineffective in maintaining adduction, may well have acted to pull the mandible forward, in back of the premaxillary beak, in the last stages of adduction. Abrasion of the sides of the inner maxillary and outer dentary teeth indicates that tooth-to-tooth contact did occur. Whether such abrasion was due to contact in simple vertical adduction or in anteroposterior sliding is impossible to determine, but the evidence considered above indicates the latter probability.

Similarities of *Protorothyris* to sphenacodont pelycosaurs in the shape of the skull and palate already commented upon by Watson (1954) and Hotton (1961) suggest that the condition of the adductors in *Dimetrodon* is a retention of the primitive reptilian pattern, with modifications mainly limited to an increase in size of the temporalis.

Captorhinus, however, seems to have departed rather radically from the primitive pattern, developing specializations of the adductors that are correlated with the flattening of the skull, the peculiar marginal and anterior dentition, the modifications of the quadrate-articular joint, and the development of the coronoid process.

Thrinaxodon

The evidence for the position and extent of the external adductors of the lower jaw in *Thrinaxodon* was secured in part from dissections of *Didelphis marsupialis*, the Virginia opossum. Moreover, comparison of the two genera reveals striking similarities in the shape and spatial relationships of the external adductors. These are compared below in some detail.

The sagittal crest in *Thrinaxodon* is present but low. It arises immediately in front of the pineal foramen from the confluence of bilateral ridges that extend posteriorly and medially from the base of the postorbital bars. The crest diverges around the foramen, reunites immediately behind it, and continues posteriorly to its junction with the supraoccipital crest (Estes, 1961).

In *Didelphis* the sagittal crest is high and dorsally convex in lateral aspect, arising posterior to and medial to the orbits, reaching its greatest height near the midpoint, and sloping down to its termination at the supraoccipital crest. Two low ridges extend posteriorly from the postorbital process to the anterior end of the sagittal crest and correspond to ridges in similar position in *Thrinaxodon*.

The supraoccipital crest flares upward to a considerable extent in *Thrinaxodon* and slopes posteriorly from the skull-roof proper. The crest extends on either side downward to its confluence with the zygomatic bar. The area of the crest that is associated with the temporal musculature is similarly shaped in *Didelphis*.

The zygomatic bar in each genus is stout, laterally compressed, and dorsally convex on both upper and lower margins. At the back of the orbit of *Thrinaxodon*, the postorbital process of the jugal extends posterodorsally. At this position in *Didelphis*, there is but a minor upward curvature of the margin of the bar.

In *Thrinaxodon* the dorsal and ventral postorbital processes, arising from the postorbital and jugal bones respectively, nearly meet but remain separate. The orbit is not completely walled off from the adductor chamber. The corresponding processes in *Didelphis* are rudimentary so that the confluence of the orbit and the adductor chamber is complete.

The adductor chamber dorsally occupies slightly less than half of the total length of the skull of *Thrinaxodon*; in *Didelphis* the dorsal length of the chamber is approximately half of the total length of the skull.

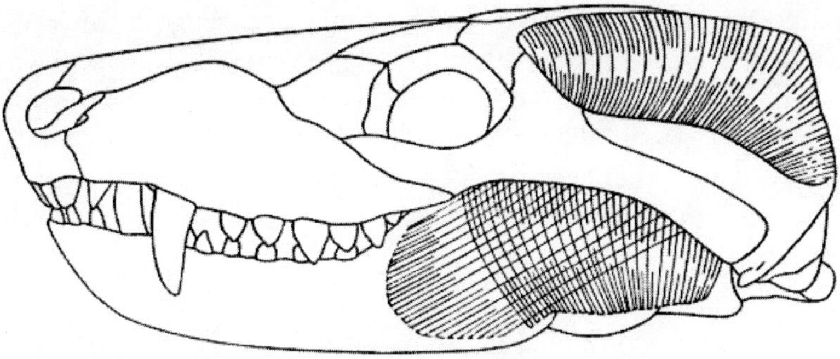

Fig. 7. Thrinaxodon. Showing masseter and temporal muscles. Skull after Romer (1956). Approx. × 7/10.

The coronoid process in *Thrinaxodon* sweeps upward posterodorsally at an angle oblique to the long axis of the ramus. Angular, surangular and articular bones extend backward beneath and medial to the process. The process extends above the most dorsal point of the zygomatic bar, as in *Didelphis*. The mandibular ramus is ventrally convex in both genera.

The relationships described above suggest that *Thrinaxodon* and the therapsids having similar morphology in the posterior region of the skull possessed a temporal adductor mass that was split into major medial and lateral components (Fig. 7). The more lateral of these, the masseter, arose from the inner surface and lower margin of the zygomatic bar and inserted on the lateral surface of the coronoid process.

The medial division or temporal arose from the sagittal crest and supraoccipital crest and the intervening dermal roof. The muscle inserted on the inner and outer surfaces of the coronoid process and possibly on the bones beneath.

Thrinaxodon represents an advance beyond *Dimetrodon* in several respects. The zygomatic bar in *Thrinaxodon* extends relatively far forward, is bowed outward and dorsally arched. Consequently, the masseter was able to extend from an anterodorsal origin to a posterior and ventral insertion. The curvature of the jaw transforms the anterodorsal pull of the muscle into a dorsally directed adductive movement regardless of the initial angle of the jaw. This is the generalized mammalian condition.

With the development of the secondary palate the area previously available for the origin of large anterior pterygoid muscles was reduced. The development of the masseter extending posteroventrally from an anterior origin presumably paralleled the reduction of the anterior pterygoids. The therapsid masseter, as an external muscle unhindered by the crowding of surrounding organs, was readily available for the many modifications that have been achieved among the mammals.

In the course of synapsid evolution leading to mammals, the temporal presumably became the main muscle mass acting in adduction of the lower jaw. Its primacy is reflected in the phyletic expansion of the temporal openings to permit greater freedom of the muscles during contraction. In the synapsids that lead to mammals, there is no similar change in the region of the palate that can be ascribed to the effect of the pterygoid musculature, even though these adductors, like the temporal, primitively were subjected to severe limitations of space.

Didelphis

Dissections reveal the following relationships of the external adductors of the jaw in *Didelphis marsupialis* (Fig. 8).

1. MASSETER

Origin: ventral surface of zygomatic arch.

Insertion: posteroventral and lateroventral surface of mandible.

2. EXTERNAL TEMPORALIS

Origin: sagittal crest; anteriorly with internal temporalis from frontal bone; posteriorly with internal temporalis from interparietal bone.

Insertion: lateral surface of coronoid process of mandible.

3. INTERNAL TEMPORALIS

Origin: sagittal crest and skull roof, including posterior two-thirds of frontal bone, whole of parietal, and dorsalmost portions of squamosal and alisphenoid.

Insertion: medial surface of coronoid process; dorsal edge of coronoid process.

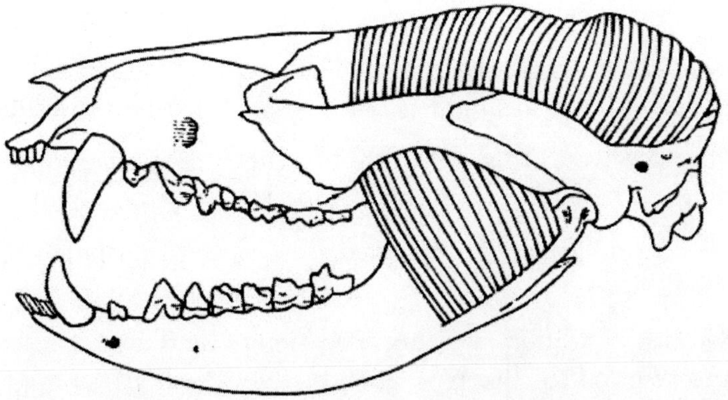

Fig. 8. Didelphis marsupialis. Showing masseter and temporal muscles. Skull KU 3780, 1 mi. N Lawrence, Douglas Co., Kansas. × 3/5.

Temporal Openings

In discussions of the morphology and functions of the adductor mechanism of the lower jaw, the problem of accounting for the appearance of temporal openings in the skull is often encountered. Two patterns of explanation have evolved. The first has been the attempt to ascribe to the constant action of the same selective force the openings from their inception in primitive members of a phyletic line to their fullest expression in terminal members. According to this theory, for example, the synapsid opening appeared *originally* to allow freer expansion of the adductor muscles of the jaw during contraction, and continued selection for that character caused the openings to expand until the ultimately derived therapsid or mammalian condition was achieved.

The second course has been the attempt to explain the appearance of temporal openings in whatever line in which they occurred by the action of the same constant selective force. According to the reasoning of this theory, temporal fenestration in all groups was due to the need to decrease the total weight of the skull, and selection in all those groups where temporal fenestration occurs was to further that end.

Both of these routes of inquiry are inadequate. If modern views of selection are applied to the problem of explaining the appearance of temporal fenestrae, the possibility cannot be ignored that:

1. Selective pressures causing the inception of temporal fenestrae differed from those causing the continued expansion of the fenestrae.

2. The selective pressures both for the inception and continued expansion of the fenestrae differed from group to group.

3. Selection perhaps involved multiple pressures operating concurrently.

4. Because of different genotypes the potential of the temporal region to respond to selective demands varied from group to group.

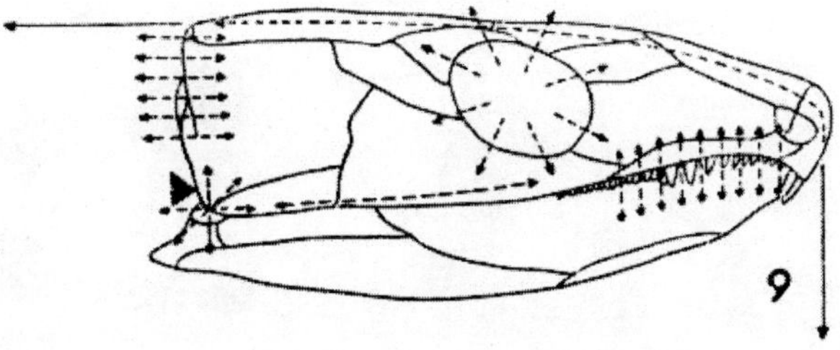

Fig. 9. Captorhinus. Diagram, showing some hypothetical lines of stress. Approx. × 1.

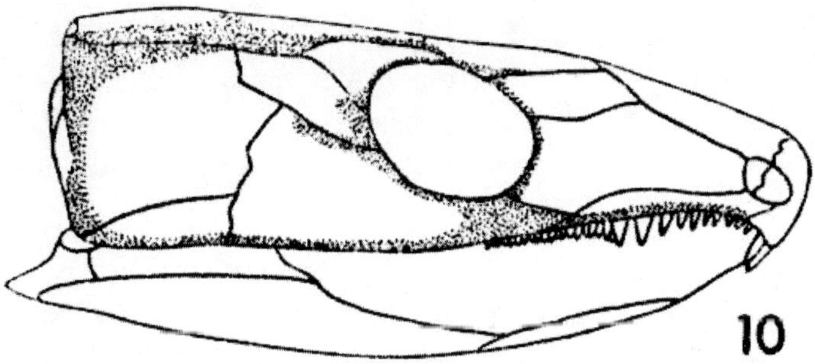

Fig. 10. Captorhinus. Diagram, showing areas of internal thickening. Approx. × 1.

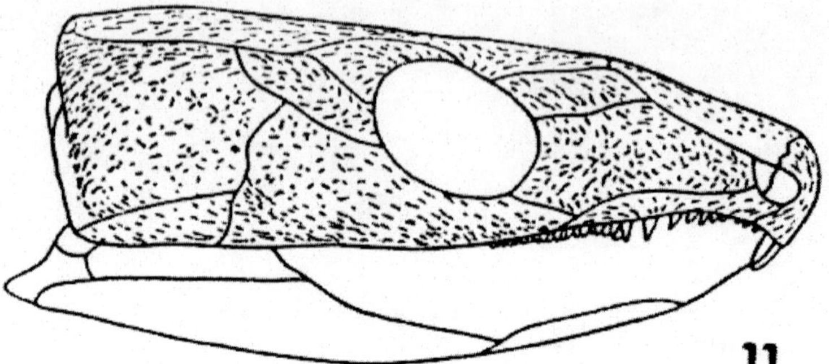

Fig. 11. Captorhinus. Diagram, showing orientation of sculpture. Approx. × 1.

Secondly, the vectors of mechanical force associated with the temporal region are complex (Fig. 9). Presumably it was toward a more efficient mechanism to withstand these that selection on the cheek region was operating. The simpler and more readily analyzed of these forces are:

1. The force exerted by the weight of the skull anterior to the cheek and the distribution of that weight depending upon, for example, the length of the snout in relation to its width, and the density of the bone.

2. The weight of the jaw pulling down on the suspensorium when the jaw is at rest and the compression against the suspensorium when the jaw is adducted; the distribution of these stresses depending upon the length and breadth of the snout, the rigidity of the anterior symphysis, and the extent of the quadrate-articular joint.

3. The magnitude and extent of the vectors of force transmitted through the occiput from the articulation with the vertebral column and from the pull of the axial musculature.

4. The downward pull on the skull-roof by the adductor muscles of the mandible.

5. The lateral push exerted against the cheek by the expansion of the mandibular adductors during contraction.

6. The necessity to compensate for the weakness in the skull caused by the orbits, particularly in those kinds of primitive tetrapods in which the orbits are large.

The distribution of these stresses is further complicated and modified by such factors as:

1. The completeness or incompleteness of the occiput and the location and extent of its attachment to the dermal roof.

2. The size and rigidity of the braincase and palate, and the extent and rigidity of their contact with the skull.

The stresses applied to the cheek fall into two groups. The first includes all of those stresses that ran through and parallel to the plane of the cheek initially. The weight of the jaw and snout, the pull of the axial musculature, and the necessity to provide firm anchorage for the teeth created stresses that acted in this manner. The second group comprises those stresses that were applied initially at an oblique angle to the cheek and not parallel to its plane. Within this group are the stresses created by the adductors of the jaw, pulling down and medially from the roof, and sometimes, during contraction, pushing out against the cheek.

It is reasonable to assume that the vectors of these stresses were concentrated at the loci of their origin. For example, the effect of the forces created by the articulation of the jaw upon the skull was concentrated at the joint between the quadrate, quadratojugal, and squamosal bones. From this relatively restricted area, the stresses radiated out over the temporal region. Similarly, the stresses transmitted by the occiput radiated over the cheek from the points of articulation of the dermal roof with the occipital plate. In both of these examples, the vectors paralleled the plane of the cheek bones. Similar radiation from a restricted area, but of a secondary nature, resulted from stresses applied obliquely to the plane of the cheek. The initial stresses caused by the adductors of the jaw resulted from muscles pulling away from the skull-roof; secondary stresses, created at the origins of these muscles, radiated out over the cheek, parallel to its plane.

The result of the summation of all of those vectors was a complex grid of intersecting lines of force passing in many directions both parallel to the plane of the cheek and at the perpendicular or at an angle oblique to the perpendicular to the plane of the cheek.

Complexities are infused into this analysis with the division of relatively undifferentiated muscles into subordinate groups. The differentiation of the muscles was related to changing food habits, increased mobility of the head, and increase in the freedom of movement of the shoulder girdle and forelimbs (Olson, 1961:214). As Olson has pointed out, this further localized the stresses to which the bone was subjected. Additional localization of stresses was created with the origin and development of tetrapods (reptiles) that were independent of an aquatic environment and were subjected to greater effects of gravity and loss of bouyancy in the migration from the aqueous environment to the environment of air. The localization of these stresses was in the border area of the cheek, away from its center.

What evidence is available to support this analysis of hypothetical forces transmitted through the fully-roofed skull of such an animal as *Captorhinus*?

It is axiomatic that bones or parts of bones that are subject to increased stress become thicker, at least in part. This occurs ontogenetically, and it occurs phylogenetically through selection. Weak bones will not be selected for. Figure 10 illustrates the pattern of the areas of the skull-roof in the temporal region that are marked on the internal surface by broad, low thickened ridges. The position of these ridges correlates well with the position of the oriented stresses that were presumably applied to the skull of *Captorhinus* during life. It can be seen from Figure 10 that the central area of the cheek is thinner than parts of the cheek that border the central area. The thickened border areas were the regions of the cheek that were subjected to greater stress than the thin central areas.

External evidence of stress may also be present. The pattern of sculpturing of *Captorhinus* is presented in Figure 11. The longer ridges are arranged in a definite pattern. Their position and direction correlates well with the thickened border of the cheek, the region in which the stresses are distinctly oriented. For example, a ridge is present on the internal surface of the squamosal along its dorsal border. Externally, the sculptured ridges are long and roughly parallel, both to each other and to the internal ridge.

The central area of the cheek is characterized by a reticulate pattern of short ridges, without apparent orientation. The thinness of the bone in this area indicates that stresses were less severe here. The random pattern of the sculpture also indicates that the stresses passed in many directions, parallel to the plane of the cheek and obliquely to that plane.

Possible Explanation for the Appearance of Temporal Openings

Bone has three primary functions: support, protection and participation in calcium metabolism. Let us assume that the requirements of calcium metabolism affect the mass of bone that is selected for, but do not grossly affect the morphology of the bones of that mass. Then selection operates to meet the needs for support within the limits that are set by the necessity to provide the protection for vital organs. After the needs for protection are satisfied, the remaining variable and the one most effective in determining the morphology of bones is selection for increased efficiency in meeting stress.

Let us also assume that bone increases in size and/or compactness in response to selection for meeting demands of increased stress, but is selected against when requirements for support are reduced or absent. Selection against bone could only be effective within the limits prescribed by the requirements for protection and calcium metabolism.

We may therefore assume that there is conservation in selection against characters having multiple functions. Since bone is an organ system that plays a multiple role in the vertebrate organism, a change in the selective pressures that affect one of the roles of bone can only be effective within the limits set by the other roles. For example, selection against bone that is no longer essential for support can occur only so long as the metabolic and protective needs of the organism provided by that character are not compromised. If a character no longer has a positive survival value and is not linked with a character that does have a positive survival value, then the metabolic demands for the development and maintenance of that character no longer have a positive survival value. A useless burden of metabolic demands is placed upon the organism because the character no longer aids the survival of the organism. If selection caused, for example, muscles to migrate away from the center of the cheek, the bone that had previously provided support for these muscles would have lost one of its functions. If in a population of such individuals, variation in the thickness of the bone of the cheek occurred, those with thinner bone in the cheek would be selected for, because less metabolic activity was diverted to building and maintaining what is now a character of reduced functional significance. A continuation of the process would eliminate the bone or part of the bone in question while increasing the metabolic efficiency of the organism. The bone is no longer essential for support, the contribution of the mass of bone to calcium metabolism and the contribution of this part of the skeleton to protection have not been compromised, and the available energy can be diverted to other needs.

The study of *Captorhinus* has indicated that the central area of the cheek was subjected to less stress than the border areas. A similar condition in basal reptiles may well have been present. A continued trend in reducing the thickness of the bone of the cheek in the manner described above may well have resulted in the appearance of the first reptiles with temporal fenestrae arising from the basal stock.

Such an explanation adequately accounts for an increased selective advantage in the step-by-step thinning of the cheek-wall prior to the time of actual breakthrough. It is difficult to see the advantage during such stages if explanations of weight reduction or bulging musculature are accepted.

After the appearance of temporal fenestrae, selection for the classical factors is quite acceptable to explain the further development of fenestration. The continued enlargement of the temporal fenestrae in the pelycosaur-therapsid lineage undoubtedly was correlated with the advantages accrued from securing greater space to allow increased lateral expansion of contracting mandibular adductors. Similarly, weight in absolute terms can reasonably be suggested to explain the dramatic fenestration in the skeletons of many large dinosaurs.

Literature Cited

ADAMS, L. A.
1919. Memoir on the phylogeny of the jaw muscles in recent and fossil vertebrates. Annals N. Y. Acad. Sci., 28:51-166, 8 pls.

ESTES, R.
1961. Cranial anatomy of the cynodont reptile *Thrinaxodon liorhinus.* Bull. Mus. Comp. Zool., 125(6):165-180, 4 figs., 2 pls.

HOTTON, N.
1960. The chorda tympani and middle ear as guides to origin and development of reptiles. Evolution, 14(2):194-211, 4 figs.

OLSON, E. C.
1961. Jaw mechanisms: rhipidistians, amphibians, reptiles. Am. Zoologist, 1(2):205-215, 7 figs.

ROMER, A. S.
1928. Vertebrate faunal horizons in the Texas Permo-Carboniferous redbeds. Univ. Texas Bull., 2801:67-108, 7 figs.
1956. Osteology of the reptiles. Univ. Chicago Press, xxii + 772 pp., 248 figs.

ROMER, A. S. and PRICE, L. I.
1940. Review of the Pelycosauria. Geol. Soc. Amer. Special Papers, No. 28, x + 538 pp., 71 figs., 46 pls.

WATSON, D. M. S.
1948. *Dicynodon* and its allies. Proc. Zool. Soc. London, 118:823-877, 20 figs., 1 pl.
1954. On *Bolosaurus* and the origin and classification of reptiles. Bull. Mus. Comp. Zool., 111(9):200-449, 37 figs.

Transmitted December 5, 1963.

30-1522

Made in United States
North Haven, CT
04 November 2026

21769089R00021